The Cultivation

of

CITRUS FRUITS.

A Short Treatise
with Special Reference to
Fertilization.

———

HILLMAN.

Second Edition, Revised and Extended.

PUBLISHED BY
WILLIAM S. MYERS, F.C.S., Director,
Nitrate of Soda Propaganda.
Late of New Jersey State Agricultural College.
John Street and 71 Nassau, New York, U. S. A.

Preface.

The following leading authorities, among others, have been consulted in the preparation of this work :

1. *California Fruits and How to Grow Them;* by Edward J. Wickson, A.M., San Francisco, 1900.

2. *Citrus Fruit Growing in the Gulf States;* by P. H. Rolfs ; Farmers' Bulletin No. 238, U. S. Department of Agriculture.

3. *Citrus Fruit Culture;* by J. W. Mills ; Bulletin No. 138, University of California, Agricultural Experiment Station, Sacramento, 1902.

4. *Twenty-second Report of the Agricultural Experiment Station of the University of California,* Sacramento, 1904.

5. *Annual Report of the Porto Rico Agricultural Experiment Station, 1904;* by D. W. May ; U. S. Department of Agriculture, Office of Experiment Stations.

6. *Propagation and Marketing of Oranges in Porto Rico;* by H. C. Henricksen ; Bulletin No. 4, Porto Rico Agricultural Experiment Station ; Washington, 1904.

7. *Citrus Fruits in Hawaii;* by J. E. Higgins; Bulletin No. 9, Hawaii Agricultural Experiment Station ; Washington, 1905.

8. *Pomelos;* by H. Harold Hume ; Bulletin No. 58, Florida Agricultural Experiment Station ; Jacksonville, Fla., 1901,

9. *El Naranjo;* por el Dr. B. Aliño ; Valencia, 1900.

10. *Tratado completo del Naranjo;* por Prof. B. Giner Aliño ; Valencia, 1901.

JOSEPH HILLMAN.

London, January, 1907.

The Cultivation of Citrus Fruits.

A Short Treatise with Special Reference to Fertilization.

The cultivation of citrus fruits forms an important Industry as well in extensive districts of California as in Florida and sections of Louisiana and Mississippi; it has established itself in Hawaii and it has passed the experimental stage in Porto Rico and elsewhere in the West Indies.

Although successful fruit-growing is a highly remunerative pursuit, much is required to insure success. Good judgment, with expert knowledge, needs to be exercised in the selection of soil and location, in the choice of stock and bud, in the cultivation and fertilization of the grove, in the treatment of diseases and in the picking, packing and marketing of the crop.

The variety of conditions prevailing in the wide regions over which citrus fruits may be cultivated with profit, extending, as these do, from about 35° North Latitude on the Pacific Coast of the United States to about 18° North Latitude, or well within the Tropics, renders it impracticable to lay down rules of unvarying application for their culture in all districts. All that can be undertaken within the limits of this brief treatise is to indicate certain principles applicable to the rational practice of horticulture, wherever exercised, and to make suggestions regarding soils, climatic conditions, selection of varieties, suitable fertilization, methods of cultivation, and the rest, that may afford general guidance.

Again, although the orange may be taken as a type representative of the other members of the group of citrus fruits in many respects, and notably in regard to their requirements in the matter of soils and cultural methods, special treatment in certain particulars is required for the profitable cultivation of lemons, limes and pomelos, and these will, accordingly, be separately touched upon.

Climatic Conditions.

Climatic conditions must be a primary consideration in the selection of a locality for fruit-growing on a commercial scale.

It may be stated generally that the winter temperature should not fall below 26° to 27° F. of continued cold, although a temperature of as low as 24° F., if not continued for more than a few hours at a time, will be withstood by orange trees when in a dormant condition.

In Florida, the danger point of cold is regarded as being 28° F. for fruit and 24° F. for foliage.

The mean temperature of seasons is of more importance than the mean temperature of the year. In other words, the relative distribution of heat over the seasons, rather than the absolute amount received during the year, is that which determines the fitness or unfitness of the climate of a district for the growing of citrus fruits.

The rainfall should not be excessive; certainly not more than 50 to 70 inches annually. Heavy rainfall is especially a disadvantage if it occurs at the time when the trees should be dormant preparatively to blooming, or at the season when the fruit has to be marketed. Thus, citrus fruit orchards should never be planted where autumn and winter rains are the rule, as on the rainy side of islands in the West Indies. The orange tree is exacting of a regulated water supply; without this it will yield poor fruit.

The site should be sufficiently elevated above the low ground of the region to secure free drainage and immunity from frosts; whilst, where irrigation must be practised, the plantation should be so located that water is easy of access and distribution.

Soils.

Although citrus trees will adapt themselves to almost any kind of soil, the orange tree in particular thrives best in a deep, moderately rich and permeable soil; one fairly retentive of moisture and yet not heavy enough to prevent escape of excessive rainfall. It should be light enough to work readily and yet not so loose as to dry out rapidly.

Sub-strata of hard-pan or of sand and gravel must be carefully avoided, and examination for such defects should be made before laying out a plantation.

A siliceous loam with some lime and clay, deep and with capacity for necessary irrigation, is that which the orange tree prefers. Its lightness and depth allow the root system of the tree to extend and develop easily, thus inducing

rapid growth; the cultivation of such soil requires but little
labour and can be carried to a sufficient depth, so that deep
rooting is promoted; whilst the injurious extremes of drought
and water-logging are avoided.

At the same time, it is undoubtedly true that whilst
rich alluvial soils produce citrus trees of luxuriant growth,
which often bear enormous crops, the finest and choicest
fruits are largely produced upon soils of a much lower grade
of fertility. As Rolfs has remarked:

"In fertile soils the plant food is seldom properly balanced and
present in the condition best suited for producing the finest fruits, nor is it
possible to influence the contents or quality of the fruit by applying different
forms of chemical fertilizers. If, therefore, a field is normally sufficiently
fertile to produce a citrus crop for an indefinite number of years, it is
usually impossible to influence the quality of fruit markedly by means of
fertilizers. Upon soils which are nearly sterile, however, trees may be
started and fed with just such chemicals as will produce the finest quality of
fruit. It therefore happens that soils which formerly were considered
absolutely worthless for agricultural purposes are now made to produce large
crops of most excellent fruit."[*]

We are not prepared to endorse to their full logical
sequence the views thus expressed; but, with certain limita-
tions, they serve to inculcate forcibly the value attaching to
the skilled employment of chemical fertilizers in the production of high-class fruits.

Selection of Varieties.

No variety of citrus is suitable for cultivation in all regions of what may be termed the citrus belt. Thus, the navel
orange, which is pre-eminently adapted for California, is
of but little value in Florida. On the other hand, certain
varieties of pomelo are of exquisite flavour when fruited in
Florida, but are not of the same excellence when grown in
California. It is essential, therefore, to test in the district
in which it is to be grown the particular variety which it is
desired to introduce.

Indeed, so much will depend upon local conditions and
market requirements, that we consider it to be inadvisable
to make specific recommendations under this head. Suffice
it to say that the choicest varieties, capable at the same time
of heavy yields, and those taking a permanently prominent

---

[*] Farmers' Bulletin No. 238, U. S. Department of Agriculture.

place in the market are, *prima facie*, those which should be selected.

Preparation of the Land.

Clearing should be thorough; everything that would interfere with good cultivation should be removed; roots should be grubbed; the ground should be levelled, and, where needed, provision should be made for drainage and irrigation.

The soil should be broken up and reduced to a fine tilth, which will permit of careful planting and staking.

In most cases it is advisable to grow a field crop the first year; better cultivation and aeration of the soil is thereby secured and any sprouting from old roots is killed out.

The orange in particular should have full possession of the soil immediately surrounding it, undisputed by grass, weeds or other trees. Its success will be indifferent under the "hole-in-the-grass" method of cultivation.

Propagation and Choice of Stocks for Budding.

The orange cannot be trusted to come true from seed, and for fruiting purposes seedlings may be regarded as unprofitable to plant. Growth from the seed is now, however, the method almost exclusively followed for the production of stocks for budding, to the exclusion of growth from cuttings or from layers, and it is by far the best.

In growing orange seedlings, good plump seed should be selected and it should never be allowed to dry. Unless it is to be sown at once, it should be mixed with moist sand for storing.

The best time for sowing is after the soil has become warmed in the spring.

The choice of seedling stocks for budding is a matter of primary importance. A deep root system and broadly extending laterals, not too near the surface, are essentials to the ideal stock.

It has been said by so eminent an authority as Wickson (California Fruits, Chapter XXX, p.356) that "the orange root is the best foundation for an orange tree, and the seed of the seedling sweet orange is the main reliance". The sweet-orange, however, would appear to be a surface-growing stock which has few deeply penetrating roots. In the Uni-

versity of California Experiment Station trials, the sour-
orange stock has been found to be decidedly hardier, and
in every way better than the sweet-orange stock. Among
its other good qualities, it is resistant to alkali.

But the pomelo is deservedly becoming the favourite
stock in southern California. Its laterals are found at a
greater depth than the laterals of the sweet orange; it produces more fibrous roots than does either of the other stocks,
and the tree is consequently a ravenous feeder. It has suc-
ceeded better at the Experiment Station of the University
of California than has the sour stock, which seems to lack
uniformity of root growth, sometimes having but few laterals, in which case the crops are small.

Pomelo seedlings are said also to have made the best
growth in the nursery.

Seedlings are grown either in boxes or in the open
ground; in either case a rich sandy loam, which will not
bake, should be secured.

The seedlings appear in about six weeks, and with good
care in weeding and in keeping sufficiently—but not ex-
cessively—moist, they will make a growth of about a foot
in the first season.

Planting out in the nursery should be done so soon as
the ground is thoroughly warmed in the spring, when the
seedlings will be about a year old.

The distance between the rows in the nursery should
be at least four feet, to allow of horse cultivation. A distance of 18 inches between the plants in the rows will permit of the roots being sacked, or otherwise protected, when
the plants are to be removed to the plantation.

In taking the seedlings from the seed bed, a few should
be lifted at a time, and it is essential that their roots be kept
shaded and moist until the ground closes on them in the
nursery row.

It is important to have an even stand in the nursery,
and weak plants should be rejected. The seedling trees
are very susceptible of injury by frost, and it is wise to give
them some sort of protection during the winter.

The young plants are usually budded after being one
or two years in the nursery, or at two to three years from the
sowing of the seed. At a convenient time in the winter,
the lower shoots and thorns are removed, so as to leave a

clear stem of about six inches for the convenience of the budder.

The best season at which to bud is about the time when the seedling is starting into vigorous growth in the spring. In general it may be said, however, that budding on good citrus stock may be done at any time of the year when the bark of the stock separates easily from the wood. This always indicates a strong flow of sap. The buds must be taken from a vigorous, healthy tree of the variety desired. Good well-matured buds only should be used; those from both the base and the tip of shoots are frequently defective.

Spring buds start into growth almost immediately, and have the benefit of the whole summer season for developing and maturing wood.

After the bud has made a good start, the top of the stock should be removed at a short distance above it, and suckers on the old stock should be continually looked for and removed. The tender shoot of the bud is protected by tying to the stub, and when the growth of the bud has become sufficiently strong to allow of its supporting itself, the old stock is smoothly sawn away and the wound covered with grafting wax, or paint.

Probably, however, a better practice is to supply supporting stakes at once, and to cut the stocks close in lopping, as, when this is done, the buds are said to make more rapid growth.

Budded trees are given one or two years' growth in the nursery and one or two years' growth on the bud, which, added to the year in the seed bed, makes them three to four years of age from the sowing of the seed before they are ready for planting out in the orchard.

To cut a bud properly is not altogether a simple operation; indeed the whole process of budding is one which requires skill and practical experience, and it would be scarcely possible to give here intelligible and adequate directions for performing it with assurance of success.

Setting Out.

The number of trees to be set out to the acre will depend on the variety selected and the character of the land. Large-growing citrus trees such as pomelos and the Bahia and Tardiff sweet oranges, should not be set closer than 100 to the acre, and on soils of a high grade of fertility 75 are enough. Smaller growing varieties, such as the mandarin group of oranges and the limes, should not be set closer than 200 trees to the acre.

In a sandy loam, rich in organic matter, the trees grow much more vigorously, and should be set farther apart. In heavy clay soils, their growth is less luxuriant, and they may be set nearer together.

On the whole, the best arrangement of the trees is that of planting in hexagons, as shown in the adjoining figure.

Trees planted in Hexagons

This method allows of fifteen per cent. more trees than setting in squares, and the ground can be worked in three different directions, It also gives better facilities for irrigation.

The orange, in common with other evergreen trees, is extremely sensitive to exposure of its roots, and for this reason special precautions have to be taken in handling the young trees in the process of transplantation. The manner of handling will depend in a great measure upon the character of the nursery soil. Sacking and balling is, no doubt, the method to be preferred, but it requires a certain degree of adhesiveness in the soil. Lifting from the nursery when the soil is too dry, exposure of roots, or careless planting, will condemn the tree to a slow and sickly growth, and often kill it outright.

The practice of reducing the top to compensate for the loss of roots in removal, is essential, but care must be taken not to carry it too far, lest subsequent growth be thereby checked.

Fertilization.

Judicious and liberal fertilization is essential to the intensive culture of citrus fruits. Cultivation, pruning and irrigation, necessary as they are to the success of the plantation, fail in their object—once the natural fertility of the land is exhausted or sensibly reduced—if the plant food absorbed from the soil be not replaced.

The removal from the soil of the constituents of a succession of crops brings about sooner or later, sterility or exhaustion of the soil from default of nitrogen, phosphoric acid and potash; and this sterility or exhaustion cannot be prevented or remedied by any system of cultivation. Field operations carried out with thoroughness do, indeed, restore to the surface soil the elements of fertility which the rains wash down into the sub-soil, they hasten their solubility and prepare them for assimilation by the plant, but they replace nothing. It is indispensible therefore to make good the deficit, and, by means other than cultivation, to restore to the soil what the crops have removed from it. On this fact is based the use of manures, and in manuring skilfully and adequately lies the secret of successful fruit growing.

Barn-yard manure used in moderate quantities will restore to the soil, to some extent, the elements removed from it by continuous cropping, but in the case of citrus fruits it will be inadequate to increase production; employed in greater quantity and to an excessive extent, its effects are rather injurious than favourable, since it gives rise to all the evils attendant upon the application of organic nitrogen in too large proportion.

Moreover, although barn-yard manure is, in a general sense, termed a complete manure, it does not, when employed alone, satisfy the requirements of citrus trees, inasmuch as the quantity of some of its constituents needs to be supplemented if heavy yields and healthy vegetation are to be maintained.

Fortunately, the rational application of manures is daily becoming better understood, and the grower is now in a position, by the aid of chemical fertilizers, to raise the yield of his plantation to a maximum, and, at the same time, to secure for his trees that vigour of growth which enables them to resist unfavourable climatic conditions and parasitic attacks.

Liberal and judicious fertilization is essential to profitable results in the culture of citrus fruits; and the grower should be the less inclined to stint outlay in the purchase of fertilizers, because he may be certain of obtaining highly remunerative returns in the yield and quality of the crop. The orange tree, in particular, is one that responds generously to generous treatment by producing fruit of the highest quality and in enormous quantity.

To manure rationally, it is necessary to take into account the nature of the soil, the composition of the plant, its food requirements, the quantity and constituents of the crop and the conditions of vegetation. Acquainted with these, we are able to determine, with an approximation to scientific accuracy, what are the elements that have to be supplied and in what proportion.

The chemical composition of the orange tree—as well as that of citrus fruits generally—is complex, and varies somewhat widely, but the following, which represents the mean of a large number of analyses, will serve as a guide in the employment of manures:

Analysis of Orange Trees.

In 100 parts.

	Nitrogen.	Phosphoric Acid.	Potash.	Lime.
Fruit	0.32	0.38	0.32	0.43
Leaves	0.70	0.10	0.32	0.71
Trunk and branches	0.70	0.43	0.58	0.80

These data clearly indicate that no one element distinctly dominates the others; for although lime is present in quantity, that base exists in abundance in most soils and may be disregarded in the preparation of formulæ of manures, sufficient of it being applied in combination with phosphoric and sulphuric acids.

In the case of trees, like those of the citrus group, whose foliage is perennial, and which are of comparatively slow growth and subjected to only limited pruning, the elements of fertility consumed in the formation of leaf and wood are relatively small in quantity, and the greater part of the plant-food assimilated is expended upon the fruit. Thus, in determining the nature and quantity of the manures to be applied, the production of fruit is the principal factor to be taken into account in arriving at the volume of plant-foods abstracted from the soil and consequently having to be replaced.

Let us suppose that an acre of orange plantation (about 100 trees) produces twelve tons of fruit. According to the analysis given above, this crop will contain:

Nitrogen . . . .   85 pounds
Phosphoric acid 102   "
Potash 85   "

It will be necessary, then, to restore these elements to the soil, in one form or another, if the trees are not to suffer from want of nourishment and to cease to produce maximum crops of good quality.

On the basis of the foregoing figures, the typical formula of chemical manures, per acre of orange trees, will be:

Nitrate of Soda . . . .   560 pounds
Superphosphate of lime (16% soluble phosphoric acid) 612   "
Sulphate of potash 170   "

Obviously, however, this general formula must not be adopted without reference to specific conditions; it must be modified to meet the requirements of each particular case, according to the nature of the soil and the state of vegetation in the plantation.

It may be well to mention here that in the opinion of some of the most experienced and successful growers of oranges in Spain, all formulæ of manures for the orange tree should contain sulphate of lime and sulphate of iron, the quantity varying according to the composition of the soil. If the land is poor in lime, only gypsum (sulphate of lime) is employed; if lime is moderately abundant, both sulphates are used, and if the soil is distinctly calcareous, only the sulphate of iron.

The formulæ which are indicated later on are based upon that given above, modified in accordance with the particular requirements of the plantation.

In the first place, the composition of the soil has to be taken into account. To ascertain this requires a delicate analysis, which only a chemist can make; but the certainty which such an analysis affords in the application of manures renders it of the first importance to the horticulturist. The cost of it will be repaid in a single season by economies following upon a close determination of the nature and quantities of the fertilizers to be employed.

In default of an analysis, however, a knowledge of the physical qualities of soils is of much utility. A clay soil may be assumed to be rich in potash and poor in phosphoric acid; a calcareous or limey soil is, on the other hand, generally rich in phosphoric acid and poor in potash; sandy soils are almost always poor in plant-foods, and soils laden with organic matter—rich in humus—contain abundant nitrogen, although it is not always in a form in which it is assimilable by plants.

The typical formula will have, therefore, to be modified, more or less, according to the nature of the soil; the elements which are deficient or abound in the latter being increased or diminished correspondingly. As a general rule, all soils are deficient in phosphates, and therefore it will almost always be safe, if not necessary, to increase the quantity of the phosphatic fertilizer indicated in the general formula.

In the case of trees of the citrus family, an excess of nitrogen induces an exuberant production of wood and leaf, at the same time that the fruit is rendered thick in rind and puffy, deficient in sweetness, wanting in aroma and with a marked tendency to rot. Moreover, a superabundance of nitrogen in the food of the plant delays maturity; and this circumstance has to be taken into account according as early or late maturity of the crop is desired.

Organic nitrogen, as found in cotton-seed meal, dried blood, guano, barn-yard manure, etc., is especially apt to bring about a soft, rapid growth, and in certain regions, especially in Florida, its continued use is found almost certainly to give rise to "die-back."

If phosphoric acid is largely in excess in the fertilizer, the fruit will generally be abundant but small, of fine flavour, and having the seeds numerous, large and of great germinating power.

When potash is superabundant relatively to the other available constituents of plant food, the tree acquires but little development; the fruit, however, is juicy, very sweet, of rich flavour and delicate aroma; the rind is fine and the seeds small and few in number.

From what has been said it will be sufficiently obvious by itself alone the part of a complete fertilizer; each is the complement of the others, each supplements the action of the others, and conjointly and supplied in suitable proportions, they bring about the desired results.

The general formula which has been indicated presupposes that the trees to which the fertilizer is to be applied are large, in full bearing, and yielding a crop of about twelve tons of oranges per acre. If the plantation be capable of a heavier yield, on account of its maturity or from the variety being a prolific one, or owing to the nature of the soil, the advantages of the site, facilities for irrigation, the climate, etc., the quantities of the several fertilizers may be judiciously increased until the limit of production is found to have been reached.

On the other hand, if the plantation is naturally one capable of producing only moderate yields, the trees of medium size and facilities for irrigation absent, the same combination of fertilizers must be employed, but the applications must be less in quantity; and in determining that quantity the skill and judgment of the grower will come in.

Superphosphate of lime, containing not less than 14 to 16 per cent. of phosphoric acid, is to be preferred as the phosphatic element in the fertilizer on account of its ready solubility and relative cheapness. In soils distinctly deficient in lime Thomas Phosphate (basic slag) may in many cases be substituted with advantage. If basic slag be used, a somewhat larger quantity of it should be applied than is prescribed in the case of the employment of superphosphate of lime.

Of potash salts, it is advisable to use only the sulphate; it being the opinion of many experienced growers that the sulphate communicates to the fruit the greater delicacy and aroma.

Exact experiments carried out during the past few years have shown that Nitrate of Soda is the best source of nitrogenous food for the orange tree. And here it must be borne in mind that Nitrogen, from whatever source derived, must be in the form of a nitrate to be assimilable by the vegetable organism. Thus, if manures containing organic nitrogen, or sulphate of ammonia—which yields ammoniacal nitrogen—be employed, the nitrogen which they contain has to undergo in the soil a natural process known as Nitrification, in connection with the action of minute organisms which convert the ammoniacal or organic nitrogen of the manures into nitric acid and nitrates. For this conversion more or less time is required, and, whilst the processes last, losses occur by the giving off of free nitrogen, which the plants are unable to utilize and which is lost in the atmosphere.

To illustrate these facts, it may be mentioned that horn shavings and Nitrate of Soda contain, in equal weights, about the same quantity of nitrogen, and nevertheless their value as fertilizers is very different. The nitrogen of the nitrate of soda is in a form in which it is immediately assimilable by plants, whilst that contained, in organic form, in the horn shavings nitrifies slowly, three or four years being required for the completion of the process. The like is the case with barn-yard manure and all other animal or vegetable manures, although the nitrification of the organic nitrogen of some of them is completed in the course of two years or less.

When we remember that 15 to 16 per cent. of the weight of nitrate of soda of ordinary commercial purity (95%) is represented by Nitrogen, and that 1 cwt. of it therefore contains as much nitrogen as ultimately becomes available from the decomposition of a ton and a half of rich barn-yard manure, the activity and rapidity of action of this fertilizer, and at the same time the control which the cultivator is able to exercise over its effects, are readily to be understood.

Another important characteristic of Nitrate of Soda is the freedom with which it permeates the soil. The roots have not to wait for nitrogenous food until they can grow down to it, neither have they to seek it immediately beneath the surface, to the encouragement of shallow rooting; and, as a consequence, plantations dressed with nitrate of soda suffer less from drought than those deriving their nitrogen from other sources.

Although the intensive cultivation of citrus fruits can only be carried on with the aid of chemical fertilizers, the application of organic manures to the plantation should not be omitted, if only as a means of maintaining the mechanical condition of the soil. Where barn-yard manure is at disposal, it should be spread over the grove lightly, so that each tree receives only a small amount. Good results are also to be obtained by ploughing under, every second or third year, a leguminous crop. Among the plants suitable for green-manuring tested at the Experiment Station of the University of California, a variety of Horse Bean, *Vicia faba*, has been found to be one of the most suitable for use in citrus orchards. All the horse beans make rapid winter growth, and growth ceases with the coming of hot weather. Thus, the crop can be ploughed under early in the spring, and the ground be subsequently thoroughly pulverized for the retention of moisture during the summer drought. Their abundant root growth is an advantage in favour of these beans for green manuring. Not only do the roots assist in opening the soil, but by their decay they add materially to the humus contents of the land.

Apart from the benefit of its mechanical action on the soil, organic manure thus applied yields up the products of its decomposition little by little, thereby steadily maintaining the food supply of the tree; and when the energies of the latter have to be brought to a maximum for the production of a heavy crop, the application of the readily assimilable chemical fertilizers produces its effect with certainty and at the required vegetative period.

Young Orange Trees.

The first necessity for these is rapid development of the trunk and foliage; the following is a suitable dressing, per acre of the plantation:

Nitrate of Soda	350 pounds
Superphosphate of lime	350 "
Sulphate of potash	100 "

The addition of the following will in many cases prove advantageous:

Sulphate of lime	150 pounds
Sulphate of iron	100 "

The application should be increased to the full dressing as the yield of fruit becomes more abundant.

Old Orange Trees.

Old orange trees contain much fixed lime in the trunk and branches, and they need to have activity imparted to the sap, foliation and florescence. The following application, per acre, will be a suitable one:

Nitrate of Soda	600 pounds
Superphosphate of lime	700 "
Sulphate of potash	150 "
Sulphate of iron	200 "

Adult Orange Trees of Sickly Vegetation.

Orange trees the produce of which is scanty and the vegetation sickly and affected by chlorosis are distinctly benefited by the application of a full dressing of nitrate of soda. The following are the constituents of a suitable fertilizer per acre of the grove:

Nitrate of Soda	700	pounds
Superphosphate of lime, or Basic slag	350	"
Sulphate of potash	200	"
Sulphate of lime	300	"
Sulphate of iron	150	"

Both with this class and with old trees, once they have recovered themselves and regained normal development, the total quantity of the fertilizers should be diminished by 15 or 20 per cent.

Trees Producing Much Wood, Foliage and Flower, but Little Fruit.

Until of late years it has been the practice to dress such trees exclusively with phosphates and potash salts, it being generally believed that nitrogenous manures increased the tendency to defective fructification.

Careful observation has led to this view being discarded.

It is rare to find an orange plantation that does not contain trees of this character, which, whatever dressing and cultivation they may receive, continue to bear scantily. Recourse is now usually had to re-grafting, or "working-over", and the abundant yields frequently obtained by this means afford proof that some other cause than food deficiency had induced the partial barrenness.

Mandarin Oranges.

This variety is a greedy feeder and requires an ample supply of manures; but, as it is of less size and yields smaller crops than the ordinary species, it requires proportionately reduced quantities of the several fertilizers. When in full bearing, mandarins should receive, per acre, fertilizers in accordance with the following formula:

Nitrate of Soda	450	pounds
Superphosphate of lime	500	"
Sulphate of potash	100	"

Nurseries of Orange Trees.

Previous to planting, there should be thoroughly incorporated with the soil, for every 100 square yards:

Well-rotted barn-yard manure .... 1000 pounds
Superphosphate of lime .... 50 "
Sulphate of potash .... 20 "

In spring, 50 pounds of nitrate of soda should be broadcasted over the same area.

In the second year, the same application may be repeated with the exception of the barn-yard manure.

For the seed plot, a heavy dressing of barn-yard manure, with the addition of a moderate application of superphosphate of lime, will suffice.

In spring, the plot should be watered frequently with a solution of one-half ounce of nitrate of soda in the gallon of water.

It should be borne in mind that increased applications of fertilizers will not necessarily produce increased yields, and that, once the limits are reached of what the plant can assimilate—and those limits are approximately represented by the quantities indicated in the formulæ given above—any surplus that may be employed will be unproductive of good and economical results.

Time and Manner of Application of Fertilizers.

If any barn-yard manure is to be applied, it should be turned under at the time of the ploughing which is usually given shortly after the crop is gathered.

The superphosphate and potash salts should be cultivated in during the dormant season of the trees, or, at latest, sometime before active vegetation commences in the spring.

The nitrate of soda should be broad-casted in three successive dressings, as follows: A third-part at the outset of vegetation in the spring; another third-part at least a fortnight before flowering commences, and the remaining third-part sometime after the fruit is well set. After each dressing, shallow cultivation and irrigation may follow with advantage.

If sulphate of lime is to be employed, it may be mixed with the last dressing of nitrate of soda; and the sulphate of iron, finely pulverized, may be put on a few days later.

If the foregoing directions are acted on, the plantation will have been adequately manured before the trees come into flower, the food materials will be fully utilized, and florescence will develop uniformly and be apt for fertilization.

Manures should not be applied during the flowering period, lest an uneven flow of sap be induced and fructification be interefered with.

Earlier or later maturity of the crop depends upon the period at which the fertilizers are applied, since the earlier the nutritive principles are taken up and assimilated, the more quickly they are accumulated in the fruit, and the earlier will its development be completed.

Accordingly, if early ripening is desired, the application of the last dressing of nitrate of soda and of the sulphate of lime and sulphate of iron should be advanced; if, on the other hand, it is wished to retard the maturity of the fruit, the application of this dressing must be correspondingly delayed.

The fertilizers should be evenly distributed over the entire area beneath the branches of the trees, with the exception of a circle of about two feet from the trunk, which should be protected by a ridge of earth from contact with the manures and water. This manner of application is essential, as the fertilizers act almost exclusively through the fine absorbent rootlets.

Irrigation.

Citrus trees require liberal supplies of moisture. The exact quantity of water necessary cannot be stated, since it will vary with the character of the soil, the distribution of the rainfall, and the care taken in its conservation in the soil.

In California irrigation is general, the number of applications varying from three to eight yearly. The best practice can be determined only by the grower himself after a study of local conditions.

Irrigation by furrows as deep and narrow as practicable has been strongly recommended by Professor Hilgard.

This method consists of running a plough to a depth of a foot, or even more, in three furrows, between the rows. When the water is applied in such furrows, a team can be driven along the dry strips of land between them, and with a harrow or other appliances the dry soil can be dragged into the wet furrows immediately after the irrigation water is turned off, and evaporation be thus lessened. The surface soil is kept comparatively dry by this method and there is nothing to attract root-growth to the surface.

Cultivation.

The main objects of cultivation, using the term in its widest sense, are two: Winter cultivation for moisture reception, and summer cultivation for moisture retention. The securing of these objects underlies the use whether of the plough or of the various kinds of harrows and cultivators.

The orange requires good, clean tillage. If weeds and grass are allowed to occupy the ground the grove will suffer. The plantation should be ploughed at least once a year, immediately after the crop is gathered.

Distinct advantage will be found in varying the depth of tillage from year to year—say eight inches, twelve inches, ten inches, fourteen inches, and then eight inches again—; by this method what is known as "plow sole", or "hard pan", a hard and impervious layer of soil which forms when the land is continuously cultivated to the same depth, will be avoided.

Tillage should follow irrigation as soon as the land is dry enough to admit of it.

Pruning.

Orange and other citrus trees, except the lemon, require little pruning after the head has been properly formed. It is of great importance that the tree be given a proper shape by judicious pruning and pinching during the first years of its orchard life. The aim should be to secure a low-headed, symmetrical tree, of upright growth, covered with a compact, but not crowded, wall of foliage. Dead twigs in the fruit-bearing area should be removed and all dead branches in the interior of the tree be cut out.

Diseases of the Orange.

The most serious diseases of the orange tree are those known as "gum disease" and "die-back".

The treatment for the former is by the use of a wash of lime, crude carbolic acid and salt. The sour orange stock is said to be practically proof against gum disease. The pomelo stock is also resistant to it.

The application of organic manures, and in particular of barn-yard manure, should be altogether avoided in the case of trees suffering from gum sickness.

With regard to "die-back" Dr. E. W. Hilgard writes:

"In almost all cases of "die-back," examination has shown some fault in the sub-soil, which puts the roots under stress. Such fault may be an underlying hardpan or impervious clay, pure and simple ; or it may be excessive wetness or dryness of the sub-strata surrounding the deeper roots ; or the rise of bottom water from below, as in cases of over-irrigation. The true "die-back" is not properly a disease, but simply the manifestation of the distress felt by the root-system underground. The first thing needful is to dig down and examine the roots, and then to relieve whatever fault may be found, if possible ; which may not always be the case. Sometimes an appearance similar to the "die-back" is caused by the roots encountering a marly stratum, which is apt to stunt the growth of the tree, causing it to put out a multitude of small, thin branches, and sometimes causing the tips to die off. For this form of the trouble there is no permanent remedy ; the trees should never have been planted in such ground, any more than in such as has shallow-lying hardpan or clay."

Pomelos.

As a tree, the pomelo most nearly resembles the orange, and its culture is virtually the same. It is a rapid grower and precocious in fruit-bearing. Like all citrus fruits, the weight and quality of its yields are to a very great extent a reflection of the care and food given to the tree. It may be stated as an indisputable fact that the grower who fertilizes heavily has the largest crop, the best fruit, and the largest profit from his plantation.

To attain the standard of excellence the fruit must have the characteristic pomelo flavour—a pleasant commingling of bitterness, sweetness and acidity. Among the large varieties, Duncan, Hall, McKinley, Pernambuco, Standard and Walters are prominent. Triumph is a good variety, as well as the Marsh, the latter being notable for the small number of its seeds. Of the smaller varieties Josselyn is probably the most characteristic.

It may be calculated that each tree in full bearing will yield ten boxes of fruit, of the average weight of eighty pounds per box.

Each tree bearing ten boxes or 800 pounds of fruit will require to be supplied, to make fruit alone, with:

Superphosphate of lime (14% soluble phosphoric acid) 2.86 pounds
Sulphate of potash (50% available potash) ... 5.86 "
Nitrate of Soda (15% nitrogen) .... 5.86 "

This is on the assumption that the constituents required for wood growth, etc., will be gathered from the soil, and no allowance is made for losses by leaching, etc. In practice, therefore, the quantity of each fertilizer will require to be amplified.

According to Bulletin No. 58 of the Florida Agricultural Experiment Station, the experience of most growers points to the use of chemical fertilizers *alone* for *all* citrus trees. The grove fruits more heavily, a better quality of fruit is obtained and the trees are maintained in a healthier condition.

Where large amounts of organic fertilizers are used, *die-back* will almost surely affect the trees, and fruit containing a large amount of *rag* and of poor shipping and keeping quality is the result.

The Lemon.

The lemon requires a practically frostless situation. Under favourable conditions the tree blooms and fruits continuously through the year. It delights in a sandy loam, but it will thrive on other soils. In southern California the lemon is profitably grown upon deep clay loams, and even upon strong red clay soils.

The prevailing stock is the orange seedling. If lemon seedlings are desired they may be grown in the same way, but the lemon on its own root will sometimes fail where, grown on the orange stock, it will thrive. The budding and planting of the lemon is carried out in the same way as in the case of the orange. The distance apart of the trees in the grove varies from twenty to twenty five feet. Greater care and attention are required to bring the lemon into good bearing form and to retain it in satisfactory shape than is the case with the orange.

The lemon responds very freely to the application of fertilizers, and the quantity of nitrogen applied may with advantage be ten or twelve per cent. greater than has been prescribed for the orange.

The Lime, the Citron, and Minor Citrus Species.

The lime is much less hardy than the lemon. It has been killed in situations where the orange and lemon have not been injured.

Limes are grown from seed, the variety usually coming true from seed. The trees are small and are frequently grown in hedge form.

The Citron, on the other hand, is quite hardy. As yet there is no considerable production in California, although experimental planting is continued with some activity.

Various minor citrus species, including the Bergamot, are grown to some extent in southern California, but chiefly for curiosity or ornament.

What has been indicated above in the case of the lemon will apply generally to the fertilization of these species.

In conclusion we would say that we do not pretend to have done more in this little work than briefly touch upon salient points in connection with specified divisions of the subject. Our object has been to aid the grower of citrus fruits by pointing to methods of culture which have been ascertained—chiefly by the labours of the Experiment Stations—to be the best, and the most likely to be pursued with advantage and prospective profit.

We expressly disclaim any attempt to lay down hard-and-fast rules or to give directions which should supersede the exercise of the planter's own judgment, or replace the indications afforded by his knowledge of the particular conditions, as respects soil and climate, in which he is working.

General Directions for
the Use of Nitrate of Soda on Staple Crops.

The use of Nitrate of Soda *alone* is never recommended, except at the rate of not more than one hundred pounds to the acre. *It may be thus safely and profitably used without other fertilizers.* It may be applied at this rate as a Top-

Dressing in the Spring of the year, as soon as vegetation begins to turn green; or, in other words, as soon as the crops begin new growth. At this rate very satisfactory results are usually obtained without the use of any other fertilizer, and the Soda residual, after the Nitrogenous Ammoniate Food of this chemical is used up by the plant, has a perceptible effect in sweetening sour land.

In most of our Grass experiments where Nitrate was used alone at the rate of but One Hundred Pounds per acre, not only was the Aftermath, or Rowen, much improved, but in the subsequent seasons, with nothing applied to the plots, a decidedly marked effect was noticed, even on old meadows. This speaks very well indeed for Nitrate of Soda not leaching out of the soil. The readily soluble elements are the readily available elements. The natural capillarity of soils doubtless is, in most instances, a powerful factor in retaining all readily soluble elements of fertility.

If this were not so, all the fertility of the world in our humid regions would, in a season or two, run into the ocean, and be permanently lost.

This is mentioned on account of certain critics having taken the trouble to object to the use of Nitrate on the grounds that it would leach away. A case is yet to be seen where the after-effect of Nitrate is not distinguishable, and, in certain cases, such effects have been most marked.

When it is desired to use a larger amount than one hundred pounds per acre of Nitrate of Soda as a Top-Dressing, or in any other way, there should be present some form of Phosphatic and Potassic Plant Food, and we recommend not less than two hundred and fifty pounds of either Acid Phosphate or fine ground Raw Rock, and two hundred and fifty pounds of some high-grade Potash Salt, preferably the Sulphate, or wood ashes in twice this quantity. *A much larger amount than one hundred pounds of Nitrate per acre, when used alone on staple crops, is generally sure to give an unprofitable and unbalanced food ration to the plant.* For Market Gardening Crops, Hops or Sugar-Beets, however, somewhat more may be used alone.

When the above amounts of Phosphatic and Potassic Fertilizers are used, as much as three hundred pounds of Nitrate of Soda may be applied with profit. In applying

Nitrate in any ration it is desirable to mix it with an equal

quantity of land plaster or fine, dry loam or sand.

If you have any reason to suspect adulteration of the Nitrate you may buy, send several pounds of it to your Experiment Station for analysis, giving date of purchase, full name and address of agent, and of the Company which the seller represents.

Generally on the Pacific Coast Nitrate may be applied as a Top-Dressing after the heavy Spring rains are over, but before crops attain much of a start.

Table Showing Prices of Nitrate of Soda on the Ammoniate Basis.

Figured on Basis of 380 Pounds Ammonia in One Ton of Nitrate of Soda.

Cost per Cwt of Nitrate	Equivalent Cost Ammonia per Ton unit	Cost per ton of Nitrate.	Cost Ammonia per lb. as Nitrate.	Equivalent Cost of Nitrogen per lb.
$2.00	$2.10	$40.00	$0.105	$0.128
2.05	2.16	41.00	0.108	0.131
2.10	2.21	42.00	0.111	0.134
2.15	2.26	43.00	0.113	0.137
2.20	2.31	44.00	0.116	0.140
2.25	2.37	45.00	0.118	0.144
2.30	2.42	46.00	0.121	0.147
2.35	2.47	47.00	0.124	0.150
2.40	2.53	48.00	0.126	0.153
2.45	2.58	49.00	0.129	0.156
2.50	2.63	50.00	0.132	0.159
2.55	2.68	51.00	0.134	0.162
2.60	2.73	52.00	0.137	0.165
2.65	2.78	53.00	0.140	0.168
2.70	2.83	54.00	0.143	0.171
2.75	2.88	55.00	0.146	0.174
2.80	2.93	56.00	0.149	0.177
2.85	2.98	57.00	0.152	0.180
2.90	3.03	58.00	0.155	0.183
2.95	3.08	59.00	0.158	0.186
3.00	3.13	60.00	0.160	0.189

This table enables one to compare commercial quotations on ammoniates with accuracy. The figures themselves are not quotations in any sense of the word, and all the figures of the table refer only to one grade of Nitrate of Soda, namely: that containing 15.65 per cent. of Nitrogen, equivalent to 19.00 per cent. of ammonia. It is prepared merely in order that purchasers may compare the price of Nitrate of Soda, which is always quoted by the hundred pounds, with other ammoniates, which are quoted by the ton unit. In the first column, therefore, are given the prices per hundred weight of Nitrate of Soda; in the second column, the equivalent price of the ammonia per ton unit; in the third column, the corresponding prices per ton; in the fourth column, the cost of the contained ammonia per pound, a figure which is always discussed, but almost never explained in Station Bulletins, and in the fifth column are given the corresponding prices of the cost of the Nitrogen per pound, a figure also much discussed, but not always explained in Bulletins. The important figures to remember are the price per hundred weight, the price per ton and the equivalent price of the ammonia in the Nitrate per *ton unit.* *The table is prepared to cover fluctuations in price running from two dollars per hundred, to three dollars per hundred; or from forty dollars to sixty dollars per ton.*

Increased Yield per Acre of Crops receiving Nitrate at the rate of 100 pounds to the Acre over those receiving none.

Barley	400 pounds of grain.
Corn	280 pounds of grain.
Oats	400 pounds of grain.
Rye	300 pounds of grain.
Wheat	300 pounds of grain.
Potatoes	3,600 pounds of tubers.
Hay	1,000 pounds, barn-cured.
Cotton	500 pounds seed-cotton.
Sugar-Beets	4,000 pounds of tubers.
Beets	4,000 pounds of tubers.
Sweet Potatoes	3,900 pounds of tubers.
Cabbages	6,100 pounds.
Carrots	7,800 pounds.
Onions	1,800 pounds.
Turnips	37 per cent.
Strawberries	200 quarts.
Asparagus	100 bunches.
Tomatoes	100 baskets.
Celery	30 per cent.

Index.